Contents

Preface

Chapter 1: Introduction

Chapter 2: Poaceae: Vegetative organs

Chapter 3: Photography

Chapter 4: Grasses Herbarium Preparation

References

Preface

Welcome to the poaceae family. The Scientific study of the grasses is called agrostology.

The first chapter gives introduction to the grasses family. The second chapter describes the Poaceae family root, shoot, leaves, inflorescence, flower and the structure of the vegetative organs. The third chapter provides some pictures of grasses. The fourth chapter provides some poaceae family herbarium.

Thanks to my family. Lastly, thankful to all the friends for their continue support and encouragement. The author invites suggestions from readers for the improvement of text.

Chapter 1

Introduction

Grass is a monocotyledon plant mainly grows in the soil. They are found all over the World. Grass is belong to the family Poaceae. Approximately 12000 grasses species are found in the world. Grasses occupy wide tracts of land and they are evenly distributed in all parts of the world. They occur in every soil, in all kinds of situations and under all climatic conditions. In certain places grasses form a leading feature of the flora. As grasses do not like shade, they are not usually abundant within the forests either as regards the number of individuals, or of species. But in open places they do very well and sometimes whole tracts become grass-lands. Then a very great portion of the actual vegetation would consist of grasses.The grasses growing in pasture land and the cereals grown all over the world are of more value to man and his domestic animals than all the other plants taken together.

Table 1. Scientific Classification of the grasses

Kingdom	Plantae
Order	Poales
Family	Poaceae

John Hendley Barnhart:

John Hendley Barnhart was born in Brooklyn, New York. He graduated in 1896 from Columbia University College of Physics and Surgeons, received degree an M.D. He first given the name Poacea.

Asteraceae and Orchidaceae families are contains approximately 32,000 and 28,000 species respectively. Fabaceae and Rubiaceae families are contains approximately 19,000 and 13,500 species respectively. So, poaceae are the fifth largest plant family.

Cricket and Football Ground Contains Grasses:

One day, i am going to Jharkhand State Cricket Association International Cricket Stadium, Ranchi. I visited this stadium to showing the match. I also seen the stadium ground contains grass. So, grasses are used in cricket ground.

Figure 1.1 **Cricket ground present in the grass (Photo Credit: Pixabay)**

When, we see the football world cup match in television. The football ground contains grasses. When we go to any international football ground we saw all grounds are surrounding the grasses. So, grasses are used in football and cricket ground.

Figure 1.2 Football ground present in the grass (Photo Credit: Pixabay)

Value of grasses:

Generally grasses are green. Grass provide us food and fooder, shelter etc. Grasses are economically important plant. Many people cultivated grasses in the field and sold in the market. Many international cricket and football stadium buy grasses for their ground.

Figure 1.3 Grass seeds (Photo Credit: Pixabay

Chapter 2

Poaceae: Vegetative Organs

General Characteristics of Poacea:

General characteristics of poacea are listed below:

1. Usally, grasses are green. They are adapted in all habitats.

2. They can grow in aquatic and terrestrial conditions.

3. They are perennial or annual.

4. Poaceeaareangiospermic plant. They do not contains tap root systems.

5. Grasses are unisexual or bisexual.

Grasses are eminently adapted to occupy completely large areas of land.

6. They are also capable of very rapid extension over large areas, on account of the production of stolons, rhizomes and the formation of adventitious roots.

The root-system:

Generally, two types of roots are found in the grasses-one is primary root and other is adventitious roots. Tap roots are not found in the grasses. Roots are arise from the nodes. Roots are absorb water and mineral nutrients from the soil. The length of the roots are up to 6 inches. The primary roots develops from the embryo. Root synthesizes two hormones namely cytokinnin, gibberellin and also controls the growth of the plant. So, roots are grown under the earth.

The Shoot System:

Shoots are growns upward. Shoots consists of stems, branches, leaves, flowers, fruits, and seeds. The stem consists of nodes and internodes. They are erect and attached on the soil.

Leaves:

Leaf is composed of a blade, sheath, ligule, auricle and collar. Leaves are originated from the nodes. The lower part and upper part of the leaf is called sheath and blade respectively. Ligule is surrounds between the sheath and the blade while auricle is found in the junction of blade and sheath. Ligule can be hairy or membrane. Leaves are alternate. There are three kinds of leaves are found in the grasses- foliage leaves, scale leaves, and prophyll.

Figure 2.1 **Grasses Leaves (Photo Credit: AnupamRajak)**

Inflorescence and Flower:

The florets are aggregate to produce inflorescence of the grasses. In grasses, the inflorescence is the spikelet. The inflorescene axis i.e rachis may be cylindrical, zigzag (Pennisetumcenchroides), flattened (Paspalumscrobiculatum), trigonous (Digitariasanguinalis), jointed (Andropogon). Flowers are two types –palea and lemma. Palea is situated in the inside bract of florets and lemma is situated in the outer bract of florets. The lemma is very larger than palea. Inflorescence consists of a spikelet (the unit of grass florets). Central spikelet axis are called rachilla.

Figure 2.2 grasses inflorescence (Photo Credit: Flickr)

Flowers are bisexual or unisexual. Grass flowers are called florets. Flowers are sessile.

Figure 2.3 Grasses inflorescence (Photo Credit: Flickr)

Structure of the Vegetative Organs:

Figure 2.4 **Green colour of grasses (Photo Credit: Pixabay)**

Structure of the Stem:

The stem of a grass consists of a mass of parenchymatous cells. Stem is covered by the epidermis. The stem is usally solid and consists of nodes and internodes. Stems are hollow, cylindrical. As already stated, stems consists of nodes and internodes.

Structure of the Root:

The root of the grasses is monocotyledonous type and protected by root caps. Roots are fibrous and absorb mineral nutrients, water from the soil.

Structure of the leaf:

Leaf is an elongated structure. Leafs are arising at the node. Leafs are consisting of sheath and blade. Leaf is covered by epidermis. Epidermis are filled with parenchymatous cells. Air cavities are found in the stomata. Presence of motor cells are many grasses.

Pollination:

When, we walk on the field or ground, grasses are gather on our feet. When birds and other animals are walk on the forests or field, grasses seeds are attached of birds, animals and dispersed to another place. Grasses are wind pollinated. They do not have scented flower to attract butterflies, insects or animals for pollination.

Chapter 3

Photography

As already stated poaceae family are found in all over the world. In india, poaceae family are found in wide varieties. I take some pictures of poaceae family in bolpursantiniketan, west Bengal. Pictures are listed below-

Figure 3.1 Grasses (Photo Credit: AnupamRajak)

Figure 3.2 Grasses species (Photo Credit: AnupamRajak)

Figure 3.3 Pennisetumpedicellatum (Photo Credit: AnupamRajak)

Figure 3.4 Bamboo tree (Photo Credit: AnupamRajak)

Figure 3.5 Bamboo sp. (Photo Credit: AnupamRajak)

Figure 3.6 **Author as Herbarium Preparation (Photo Credit: AnupamRajak)**

Figure 3.9 Grasses (Photo Credit: Pixabay)

Figure 3.11 Grasses (Photo Credit: Pixabay)

Figure 3.11 Ornamental grasses (Photo Credit: Pixabay)

Chapter 4

Grasses Herbarium Preparation

What is Herbarium?

Herbarium is the collection of plant materials. After collection, the plants are pressed. Then, after few days later, plants are dried. Dried plants are attached on a sheet of high quality paper.

Location:

The authors are collected the plants in Bolpur-Santiniketan area. After collection, authors are identified grass specimens with the help of internet and research schlors.

Equipments:

When you collect plant specimens, you must be brings trowel (shovel), field press, bags, GPS tracker, Copy etc.

Grasses Distribution:

Leptochloadigitata:

Leptochloadigitata is a widespread family present in Asia, Africa. Raceme inflorescence are found in leptochloadigitat.

DigitariaCilliaris:

Digitariacilliaria is found many parts of the country. Digitariacillaris inflorescence is raceme type.

Aristidasetaceae:

Aristidasetaceae is common name is broom grass. Aristidasetaceae is found in india, srilanka, and Myanmar.

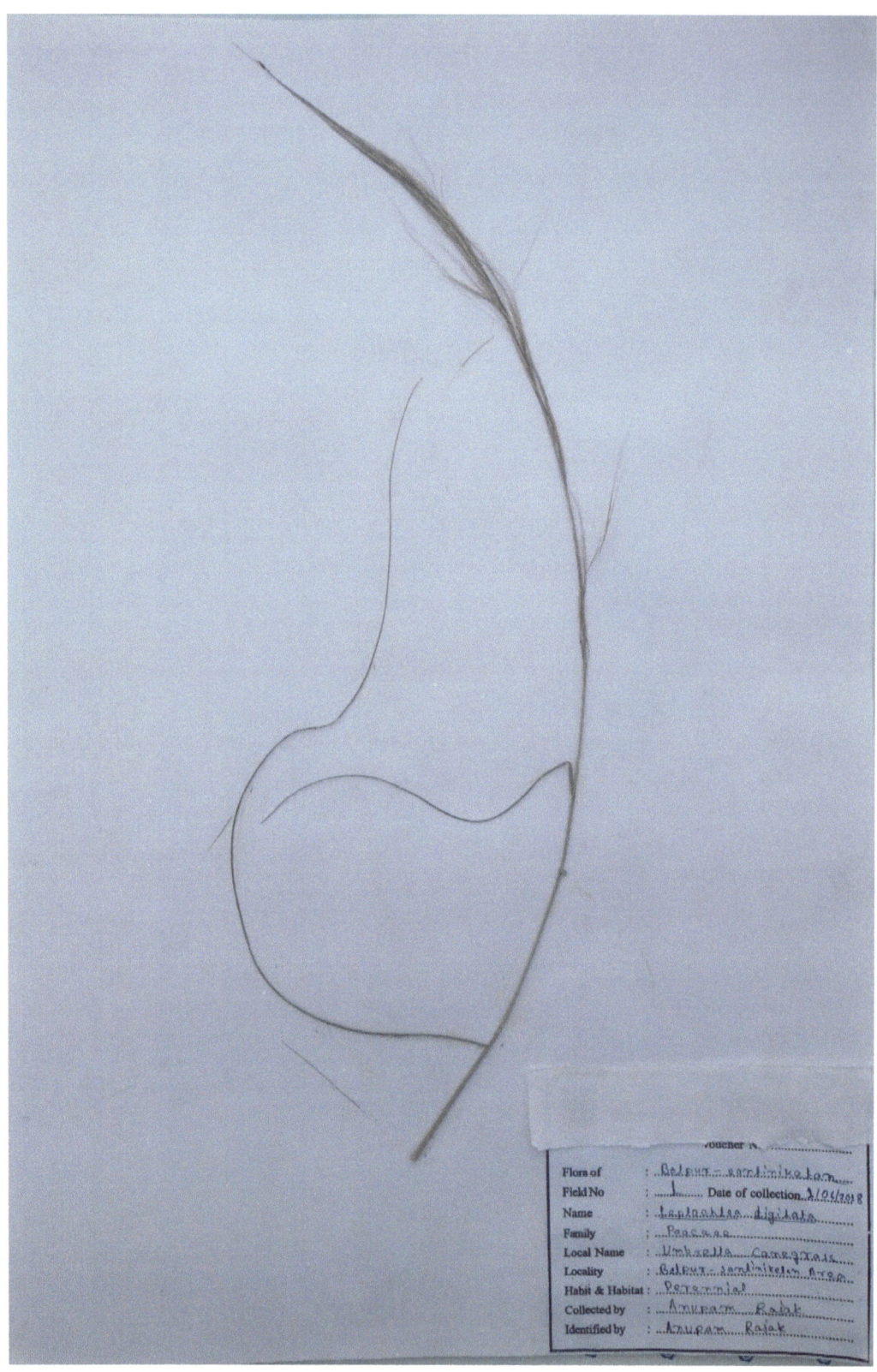

Figure 4.1 Herbarium of Leptochloa digitata (Photo credit: Anupam Rajak)

Figure 4.2 Herbarium of Digitariacilliaris (Photo credit: AnupamRajak)

Figure 4.3 Herbarium of Aristidasetaceae (Photo credit: AnupamRajak)

Figure 4.4 Herbarium of Eragrostiscoarctata (Photo credit: AnupamRajak)

Figure 4.5 Herbarium of Triticum aestivum (Photo credit: Anupam Rajak)

Figure 4.6 Herbarium of Dichanthium annulatum (Photo credit: Anupam Rajak)

Figure 4.7 Herbarium of Eragrostiscurvula (Photo credit: AnupamRajak)

Figure 4.8 Herbarium of Cynodondactylon (Photo credit: AnupamRajak)

Figure 4.9 Herbarium of Echinochloa colona (Photo credit: Anupam Rajak)

Figure 4.10 Herbarium of Chlorisbarbata (Photo credit: AnupamRajak)

Figure 4.11 Herbarium of Sporobolus virginicus (Photo credit: Anupam Rajak)

Figure 4.12 Herbarium of Oryzae sp. (Photo credit: Anupam Rajak)

Figure 4.13 **Herbarium of Eragrostispilosa(Photo credit: AnupamRajak)**

Figure 4.14 Herbarium of Brachiaria plantaginea (Photo credit: Anupam Rajak)

Figure 4.15 Herbarium of Oplismenus burmanii (Photo credit: Anupam Rajak)

Figure 4.16 Herbarium of Oplismenuscompositus(Photo credit: AnupamRajak)

Figure 4.17 Herbarium of Pennisetum pedicellatum (Photo credit: Anupam Rajak)

Figure 4.18 Herbarium of Sporobolusdiander(Photo credit: AnupamRajak)

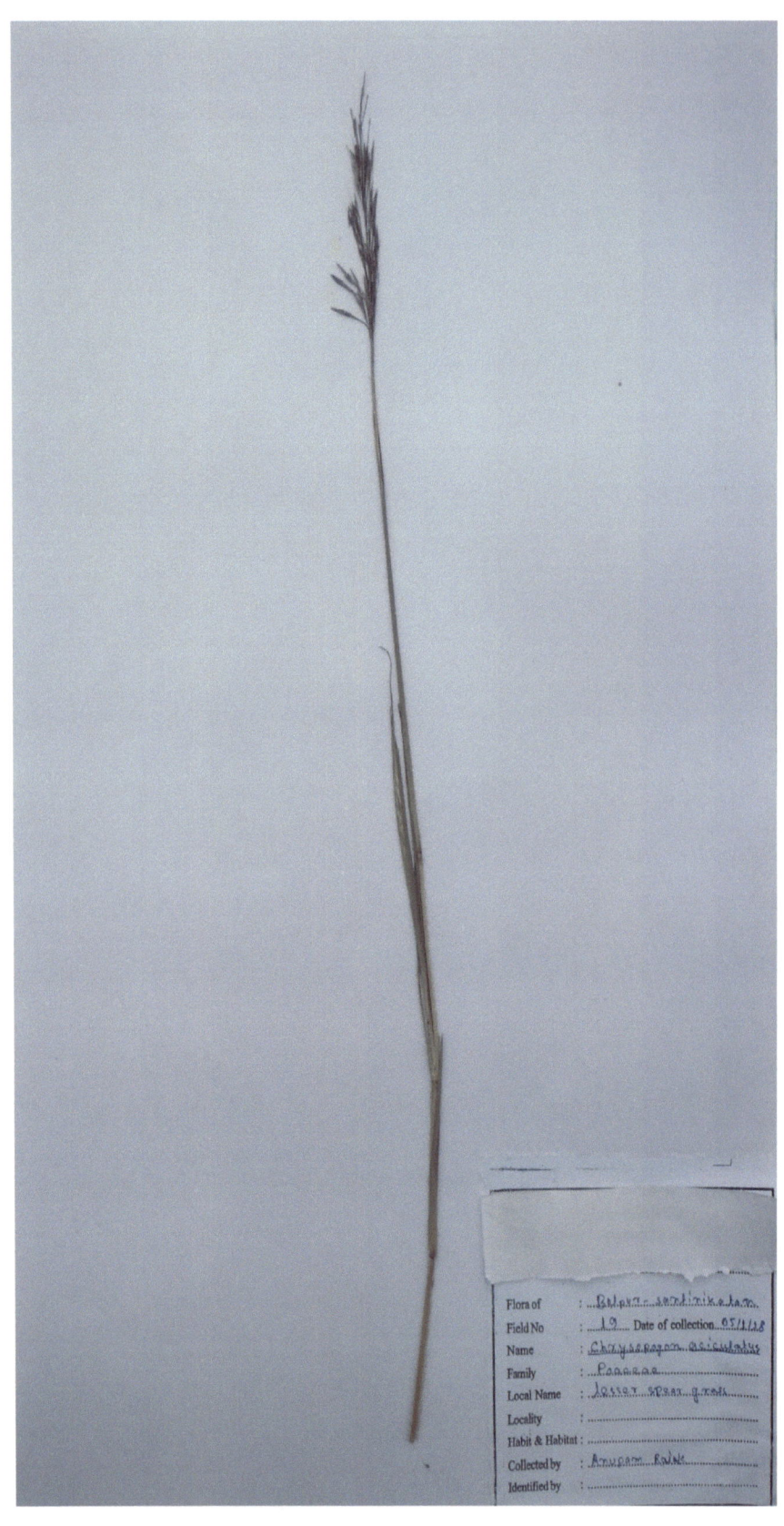

Figure 4.19 Herbarium of Chrysopogonaciculatus(Photo credit: AnupamRajak)

Figure 4.20 **Herbarium of Eragrostis sp.(Photo credit: AnupamRajak)**

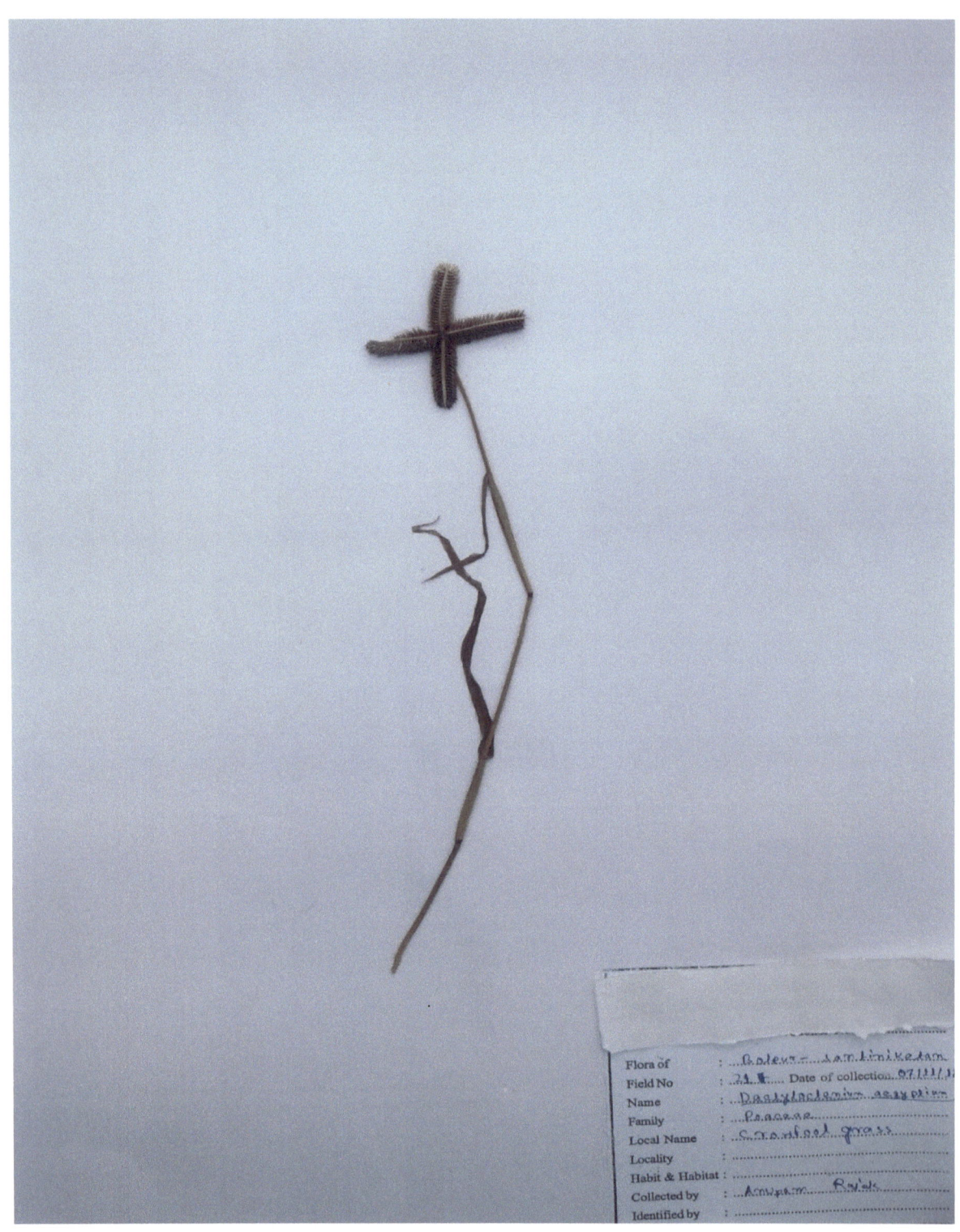

Figure 4.21 Herbarium of Dactylocteniumaegyptium (Photo credit: AnupamRajak)

Importance of grasses:

1. Poaceae family are medicinally important plants.

2. We used this plants for food.

3. Animals are used this plants for food and fooder.

4. In india, many peoples are uses grasses in puja rituals.

5. They are ecologically important plants. Grasses are used in soil conservation. They also protect soil.

6. Paper is made from bamboo.

7. Beer is also made from grasses species such as barley.

8. Sugar produced from grass species such as sugarcane.

9. Bamboos are eaten by many peoples because they are very tasty.

References:

1. BAHADUR, RAI. ACHARIYAR, K. RANGA. MUDALIYAR, C. TADULINGA. *HANDBOOK OF SOME SOUTH INDIAN GRASSES*.ALPHA EDITIONS, 2018.

2. Achariyar, RaiBahadur K. Ranga., and C. TadulingaMudaliyar. *A Handbook of Some Indian Grasses*.M/s Bishen Singh Mahendra Pal Singh.

3. Harris, Tom. "How Grass Works." *HowStuffWorks*, HowStuffWorks, 30 Apr. 2002,

4. Wikipedia

5.britannica

www.ingramcontent.com/pod-product-compliance
Lightning Source LLC
Chambersburg PA
CBHW040415220526
45473CB00004B/1251